SAE EDGE™
RESEARCH REPORT

Unsettled Technology Domains in Robotics for Automation in Aerospace Manufacturing

Dr. Jody E. Muelaner
Muelaner Engineering Ltd.

EDGE DEVELOPMENT TEAM

Dave Burns, *Kuka Systems*
Andy Francis, *GKN Aerospace*
Russell DeVlieg, *ElectroImpact*
Roger Holden, *True Position Robotics*
Jim Heley, *Advanced Aerospace Assembly Ltd*
Andy Silcox, *AMRC, Broughton*
Zheng Wang, *University of Bath*

SAE INTERNATIONAL®
Warrendale, Pennsylvania, USA

About the Publisher

SAE International® is a global association of more than 128,000 engineers and related technical experts in the aerospace, automotive, and commercial-vehicle industries. Our core competencies are lifelong learning and voluntary consensus standards development. Visit sae.org

SAE EDGE™ Research Report Disclaimer

SAE EDGE™ Research Reports focus on topics that are dynamic, in which knowledge is incomplete, and which have yet to be standardized. They represent the collective wisdom of a group of experts and serve as a practical guide to the reader in understanding unsettled subject matter. They are not meant to provide a recommended practice or protocol. The experts engaged have contributed their own thoughts and points of view, and these are not the positions of the institutions or businesses with which they are affiliated. A professional writer has collectivized their input; there is no one contributor's perspective being advanced, but rather that of a community of practitioners. SAE EDGE™ Research Reports are the property of SAE International, and SAE alone is responsible for their content.

About This Publication

SAE EDGE™ Research Reports provide state-of-the-art and state-of-the-industry examinations of the most significant topics in mobility engineering. SAE EDGE™ contributors are experts from research, academia, and industry who have come together to explore and define the most critical advancements, challenges, and future direction in areas such as vehicle automation, unmanned aircraft, Internet of Things and connectivity, cybersecurity, advanced propulsion, and advanced manufacturing.

Related Resources

SAE MOBILUS® Advanced Manufacturing Knowledge Hub
https://saemobilus.sae.org/advanced-manufacturing/

SAE Team

Frank Menchaca, Chief Product Officer
Michael Thompson, Director, Standards, Information and Research Publications
Monica Nogueira, Acquisitions Director
Jill Leonard, Product Manager
William Kucinski, Managing Technical Editor

Printed in USA

ISSN 2640-3536
e-ISSN 2640-3544
ISBN 978-1-4686-0123-7

To purchase bulk quantities, please contact: SAE Customer Service

E-mail: CustomerService@sae.org
Phone: 877-606-7323 (*inside USA and Canada*)
724-776-4970 (*outside USA*)
Fax: 724-776-0790

Visit the SAE International Bookstore at books.sae.org

About the Author

Dr. Jody E. Muelaner is a chartered mechanical engineer with a background in metrology, aerospace manufacturing, and machine design. He now specializes in writing about technical topics in a way that the target audience can easily understand.

His writing has included technical reports for Rolls-Royce and Airbus, peer-reviewed journals, and UK government reports, as well as magazines and websites. He has published several hundred articles and was awarded the Sage Best Paper Award in 2010.

Starting out in machine design, Dr. Muelaner initially worked on sawmills, waste-processing machinery, domestic appliances, and medical devices. After moving into metrology, his research focused on modeling and optimizing uncertainty in manufacturing systems, enabling right-first-time assembly, and the design of innovative laser instruments. He founded Muelaner Engineering Ltd in 2018 to provide consultancy and technical writing within advanced manufacturing and machine design.

contents

Atosan/Shutterstock.com

Unsettled Technology Domains in Robotics for Automation in Aerospace Manufacturing

Abstract

Cost reduction and increasing production rates are driving automation of aerospace manufacturing. Articulated serial robots may replace bespoke gantry automation or human operations. Improved accuracy is key to enabling operations such as machining, additive manufacturing (AM), composite fabrication, drilling, automated program development, and inspection. New accuracy standards are needed to enable process-relevant comparisons between robotic systems.

Accuracy can be improved through calibration of kinematic and joint stiffness parameters, joint output encoders, adaptive control that compensates for thermal expansion, and feedforward control that compensates for hysteresis and external loads. The impact of datuming could also be significantly reduced through modeling and optimization. Highly dynamic end effectors compensate high-frequency disturbances using inertial sensors and reaction masses. Global measurement feedback is a high-accuracy turnkey solution, but it is costly and has limited capability to compensate dynamic errors. Local measurement feedback is a mature, affordable, and highly accurate technology where the robot is required to position or align relative to some local feature. Locally clamped machine tools are an alternative approach that can utilize the flexibility of industrial robots while also enabling high-quality machined surfaces. Hybrid high-accuracy control strategies will be required for many processes.

NOTE: SAE EDGE™ Research Reports are intended to identify and illuminate key issues in emerging, but still unsettled, technologies of interest to the mobility industry. The goal of SAE EDGE™ Research Reports is to stimulate discussion and work in the hope of promoting and speeding resolution of identified issues. SAE EDGE™ Research Reports are not intended to resolve the issues they identify or close any topic to further scrutiny.

DR. JODY E. MUELANER
Muelaner Engineering Ltd.

Edge Development Team

Dave Burns, *Kuka Systems*
Andy Francis, *GKN Aerospace*
Russell DeVlieg, *ElectroImpact*
Roger Holden, *True Position Robotics*
Jim Heley, *Advanced Aerospace Assembly Ltd*
Andy Silcox, *AMRC, Broughton*
Zheng Wang, *University of Bath*

ISSN 2640-3536

Introduction

Interest in automating aerospace manufacturing using industrial robots is driven by business objectives such as increasing production rate, reducing machine downtime, increasing flexibility, and reducing costs. Articulated serial robots may be used to replace both bespoke gantry automation systems and human operations. They may also be used in cooperative operations where humans are not replaced, but robotically assisted to increase productivity.

State of the Industry

There is a rapidly growing demand for single-aisle aircraft, and this is a major driver for automation. For example, over the last 10 years, Airbus has increased A320 production volume from 34 aircraft per month to 60. The company expects this ramp-up to continue [1] with insiders talking of a rate of 100 per month. Boeing has been increasing B737 production to similar rates, although it has slowed slightly after the recent global grounding of B737 MAX [2].

A major driver for aerospace manufacturing is improving performance by achieving weight reduction and increasing the accuracy of aerodynamic surfaces. Depreciation accounts for approximately 12% of airlines' operating costs, compared with 36% for fuel [3]. Aircraft operators are, therefore, more interested in an aircraft's performance than its cost. This means that design for manufacture (DfM) is less developed than in more price-sensitive industries such as automotive, presenting a number of challenges for automation. However, reduced manufacturing cost still has a significant role to play in increasing profits for aerospace manufacturers. Without some DfM effort, the full benefits of automation will not be realized.

Industrial robots have a number of advantages when compared to bespoke gantry-based automation. Large gantry machines are highly accurate and reliable. However, they require deep foundations that increase cost and lead time while making them very difficult to move. When they do develop a fault, it is often a single point of failure that can halt production for weeks while repairs are made. Although gantries are able to operate at relatively high speeds, they are typically limited to a single spindle. Also, they operate in a completely closed cell with no possibility of humans performing concurrent operations within the cell, which severely limits throughput. Robots on the other hand, are smaller, lighter off-the-shelf machines. It is possible for multiple robots to operate simultaneously on a large aerostructure with humans also working concurrently around them. When a robot develops a fault, it can be swapped out for a replacement relatively easily, and production can continue while the robot is repaired remotely. Flexibility is also often stated as an advantage of industrial robots, although many in the industry feel that this is not a primary driver. Tier-1 suppliers have contracts with primes, such as Boeing and Airbus, that last the life of the product, and frequent design changes are not possible due to the effort required for recertification. Also, once robots are integrated into large systems involving metrology and linear motion, the cost is often high (or higher) than a gantry system.

There is a general perception that replacing humans with robots will increase production rate. However, this is not necessarily true. Many high-volume and highly optimized manufacturing systems still have many processes carried out by humans, with automotive and consumer electronics industries being good examples. The key is to automate the right things and ensure that the humans and the automated systems work together efficiently. While there are several advantages to automation (e.g., robots are more easily monitored, more consistent, and take less holiday time than humans), there is not really an appetite within the aerospace industry to replace human workers. The focus is much more on making jobs easier and safer by reducing exposure to vibration, power tools, harmful substances, and heavy lifting. Processes where operators may influence variability may also be automated to improve certification; composite layup is a prime example of this. Difficulties in recruiting and retaining skilled staff may also mean that new processes are increasingly automated. Another prime area for automation is logistical operations involving retrieving parts and tools from storage and delivering them to the work cell.

This report focuses primarily on the use of articulated robots, which have rotary joints arranged in series, to replace operations requiring high accuracy that are normally carried out by either high-accuracy automation or human workers. There will also be some consideration of parallel kinematic machines (PKMs) and automated guided vehicles (AGVs).

Improving accuracy is a generic requirement that will enable operations such as machining, AM, composite fabrication, drilling, automated program development, and inspection. It is perhaps the single biggest challenge preventing widespread adoption of robots in aerospace manufacturing. The underlying problem is that industrial robots were not designed for these types of operations. This is further complicated by the fact that many accuracy requirements come from aircraft designed 20 years ago with poorly documented design rationale. There is a general feeling within manufacturing that many tolerances may be unnecessarily demanding, but the consensus is that it's safer to keep them. However, the need for improved aerodynamics will drive a need for increased accuracy in certain areas of future aircraft; therefore increased accuracy may become necessary. Most industrial robots were designed for automotive assembly, welding, and painting. These operations only need millimeter-level accuracy. There currently isn't a large market for high-accuracy robots; although if a standard aerospace robot could prove suitable for the range of operations proposed, then this could change.

The potential applications, together with unresolved areas that offer potential solutions for improved accuracy, are summarized as a matrix in Figure 1. Aerospace structures have extremely high complexity, meaning that robots

FIGURE 1. Potential applications for high-accuracy aerospace robots, together with the unresolved issues that may offer solutions for these applications.

Unresolved areas that may provide solutions:

Potential applications:	Defining accuracy	Kinematic calibration	Datuming	Adaptive control	Reducing cutting disturbances	Feed forward control	Global measurement feedback	Local measurement feedback	Factory-wide positioning	Locally clamped machine tools	High-dynamic end-effectors	Hybrid high-accuracy control
Drilling and Filling	✖	✖	✖	✖	✖	✖	✖	✖		✖	✖	✖
Machining	✖	✖	✖	✖	✖	✖	✖	✖		✖	✖	✖
Wet operations	✖	✖	✖			✖		✖				
Inspection	✖	✖	✖	✖			✖					
Composite fabrications	✖	✖	✖	✖		✖	✖	✖				✖
Additive Manufacturing	✖	✖		✖		✖	✖					✖
Friction stir welding	✖	✖	✖	✖		✖		✖			✖	✖
Logistics									✖			

performing any of these operations will need very complex programs, unlike the more repetitive work carried out in many other industries. This means that manual teaching of robots is not feasible, and therefore absolute accuracy, rather than simply repeatability, is required in each type of operation.

A final generic unresolved area, not related to any specific application, is the need for improved human-robot collaboration.

Drilling and Filling

There is interest in using robots to replace gantry machines for drilling holes and installing fasteners in airframes. This involves a number of unresolved challenges, most notably the requirement for more accurate robots, in terms of both hole position and hole quality which requires the drilling tool to be accurately translated along the hole axis. Positional accuracy of approximately 0.2 millimeters is typically required, although the path straightness may need to be significantly better than this to achieve the required form tolerances for holes.

Because of the dynamic nature of drilling, the robot must be able to accurately follow a path while subjected to disturbances such as vibration and tool reaction forces. The optimization of cutting parameters, such as spindle speed, cutter type, feed rate, number of passes (one shot or two shots), and coolant, is a key area of research aimed at minimizing these disturbances, although this work is often proprietary. The impact of disturbances on dynamic accuracy is also strongly affected by the stiffness of the robot. Because of the way that articulated robots use rotary joints to produce all motions, even straight-line motion, joint reversals can occur at unintuitive stages of a motion path. Backlash and other hysteresis effects can, therefore, have a significant impact on dynamic accuracy. Achieving practical multifunctional end effectors that are able to both drill holes and install fasteners can be challenging.

Despite remaining challenges, state-of-the-art production facilities are demonstrating that it is possible to drill holes in large aerospace structures to the required 0.2 millimeters tolerance. An example of this is the recently opened Airbus A320 fuselage assembly plant in Hamburg, Germany. This uses crawling robots to position locally clamped machine tools, known as *Flextrack* robots, to join curved panels longitudinally, forming the cylindrical sections of the fuselage. It then uses seven-axis articulated robots, integrated with linear slideways, to create the orbital joints that connect these sections (Figure 2).

Machining

Metallic part production by machining from billet is optimally performed by dedicated computer numerical control (CNC) machine tools, and there is little scope for robots to compete in this application. CNC machines offer high stiffness and high accuracy. However, there are many other

FIGURE 2. **Airbus A320 fuselage assembly, opened October 2019; image courtesy of Airbus.**

machining operations where robots may prove suitable, for example, machining component interfaces to achieve form and fit tolerances during assembly, known as fettling. Other examples might include trimming castings and face-machining additively manufactured blanks. The challenges and potential solutions for robotic machining are closely related to those for drilling and filling.

Wet Operations (Sealant Application)

Wet operations involve the application of sealant, liquid shim, or adhesive. The wet material normally needs to be deposited at a specified distance from the edge of a component. These processes appear relatively undemanding for robots as tolerances at the millimeter scale should be within the capability of standard robots. However, component variability and datuming issues can still present challenges. Increasing component accuracy would cause an unacceptable increase in manufacturing cost. A potential solution is offered by localized measurement systems that allow robots to directly follow the edge profile.

Inspection

Certification requirements for aerospace structures mean that inspection represents a very significant part of overall manufacturing cost. Automation of these operations could, therefore, have a major impact. One example of a highly repetitive inspection task that could potentially be carried out by a robot would be fastener flushness measurement. This involves measuring the height that each countersunk fastener protrudes above the surface of the panel it is fastening.

Composite Fabrication

The layup of composites is increasingly being carried out by robots using methods such as automated tape laying and automated fiber placement. Accuracy is a key concern, in particular with respect to gaps or overlap between successive passes of the deposition head.

Automated composite fabrication machines consist of two systems: a large-scale positioning system and a local head for tow placement, trimming, or pinning. Large-scale positioning may be carried out by an articulated robot or a Cartesian machine. Verification systems may operate on one of these systems independently or may verify the global position of the tow. Local verification of tow placement, relative to the positioning system, may be carried out using a two-dimensional (2D) or three-dimensional (3D) imaging system combined with automated edge detection algorithms. This system may then also be used to measure a reference artifact or targets, enabling verification of the positioning system.

AM

Directed Energy Deposition (DED) is a type of AM that feeds build material directly into a melt pool created by an energy source. Most commonly this involves a wire fed into a plasma arc, very similar to gas-shielded arc welding. DED was initially developed using articulated robots to position the deposition head.

To perform DED using an articulated robot, three things must be supplied to the end effector: high-voltage electricity, shielding gas, and wire feedstock. Feeding the wire at a consistent speed while the robot joints are rapidly articulating presented significant process control challenges. For this reason, there is a move away from the use of articulated robots toward dedicated machines designed to produce a specific component geometry.

Friction Stir Welding

Friction stir welding is potentially a lighter, stronger, and lower cost method of joining aluminum and titanium airframe components, when compared to conventional drilling and fastening. It is well suited to robotic automation although significant reaction forces complicate this somewhat.

Logistics

The adoption of AGVs within aerospace manufacturing has been relatively slow. Robotic solutions already in use for warehouse automation could offer significant benefits for logistics within aerospace manufacturing. Companies such as Amazon have led the development of this technology. The use of overhead rail systems to move large components while maintaining an open floor can also be very effective, as can pneumatic tube transport systems for the delivery of small components. These systems have been effectively implemented in the recently opened Airbus A320 fuselage assembly plant in Hamburg, Germany (Figure 3).

FIGURE 3. **Overhead rail system with junctions, points, and turntables to direct panels around the Airbus A320 assembly plant in Hamburg; image courtesy of Airbus.**

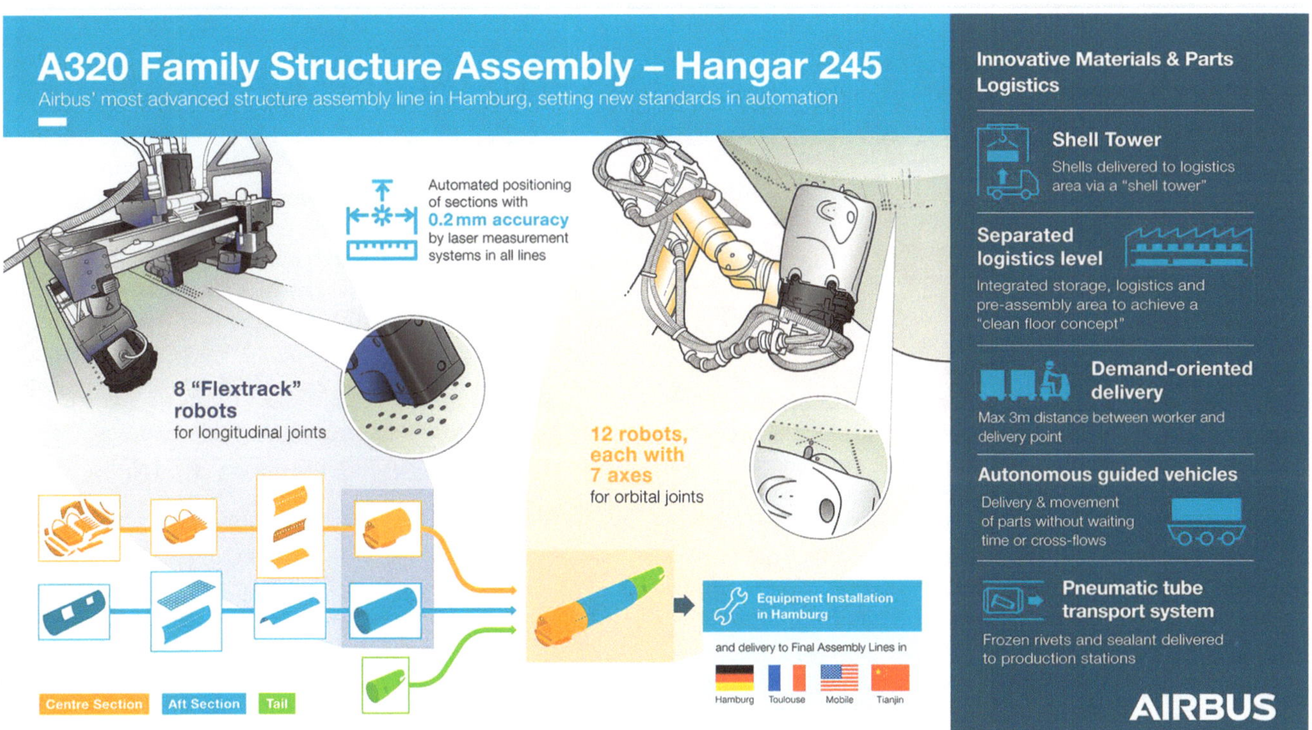

There are few unresolved areas for robotic development required to realize these benefits. The challenge lies in the process planning and logistics systems used within aerospace production. This is outside the scope of this report, but it is, nonetheless, worth noting that significant efficiency gains could be achieved through this technology.

A wide range of measurements have been used to enable AGVs to navigate within a factory environment. These include light sensors that follow lines on the floor, guide wires, vision cameras, inertial dead reckoning, indoor global positioning system (iGPS), and ultrasonic beacons.

iGPS was seen as providing a factory-wide measurement infrastructure that could be used for both low-accuracy demands (such as AGV guidance) and high-accuracy applications (such as robot control and assembly alignment.) However, it has proven too expensive for many low-accuracy applications, while having insufficient accuracy to compete with laser trackers [4]. Radio-frequency positioning systems can achieve sub-millimeter accuracy at distances of over 100 meters from transmitters [5]. A number of ultrasound time-of-flight positioning systems have been created for factory-wide positioning. Most notably, Cricket nodes are tiny open-source devices developed by the Massachusetts Institute of Technology Laboratory for Computer Science [6] and used within many research projects. They are accurate to a few centimeters within 10 meters of a transmitter [7].

Defining Accuracy

The first unresolved area, which presents a challenge for achieving the required robotic accuracy, is how to define accuracy. Different types of process require different types of accuracy, primarily in terms of its dynamics and scale.

ISO 9283 [8] defines many different ways to assess the accuracy of a robot. It states that testing should be carried out at 20°C (±2°C). All tests except for static compliance should be carried out with 100% of rated load in terms of mass, center of gravity position, and moments of inertia. All pose characteristics should be tested with the robot moving between poses at maximum speed. Perhaps the most fundamental metric is *pose accuracy*, which is defined as the difference between the commanded pose and the mean of 30 measured poses, and may be divided into positioning accuracy and orientation accuracy. This is often referred to as *absolute accuracy* to ensure that it is not confused with *pose repeatability*, defined as the closeness of agreement between *n* poses, approached in the same direction. *Multidirectional pose accuracy variation* is the repeatability when approaching from three different directions. Accuracy and repeatability may also be measured as the ability to translate by a commanded distance. *Position stabilization time* is the time that elapses from when a robot, commanded to a static position, enters the specified repeatability zone around the nominal position to when it no longer leaves the zone. *Path accuracy* is the bias in the path compared to the commanded path and must be considered with *path repeatability* to give a complete picture of accuracy. Path accuracy is important for dynamic operations but depends on many factors including velocity, minimum path radius, and robot pose. It is also very difficult to measure accurately. *Static compliance* is specified in millimeter per N in each of the axes of the base coordinate system, for both the positive and negative directions. Measurements are made with servos on and brakes off. Many other types of accuracy are defined including *position overshoot*, *cornering deviations*, path velocity characteristics, and *weaving deviations*. Standard methods of evaluating robot accuracy should, in principle, make comparisons straightforward. However, many methods are defined, and yet still none may give an accurate reflection of actual performance for a particular process. For example, path accuracy provides a good indication of machining performance, although the ultimate test is the accuracy of machined surfaces since cutting forces and vibration, as well as tool errors, will also affect machining accuracy. Therefore process-specific tests are frequently used, and literature often reports the results of these tests. As a consequence, comparisons between these studies are difficult.

The wide range of accuracy definitions provided by ISO 9283 causes significant confusion while not providing a framework that enables precise specification of a robot for a particular task. It is common for application-specific tests to be used within both industrial trials and academic studies. Tests might involve measuring positional or path following accuracy with a laser tracker [9, 10], moving over a reference artifact with a proximity sensor, drilling holes in coupons, ballbar tests [11, 12], or machining coupons.

The wide variety of tests used within studies makes the comparison of accuracy achieved by different methods of accuracy improvement very difficult.

Recommendations

There is a need to define standards for robot accuracy that enable comparison between both different commercially available robots and experimental results. Such a standard must consider the variables such as the scale over which the accuracy was measured, the dynamic nature of the accuracy, environmental disturbances, such as temperature, that were compensated, and force disturbances that were present.

To enable ready comparisons, a few sets of variable values should be used to represent conditions for processes such as

- Static positioning
- Drilling, machining, or friction stir welding
- Sealant application or dynamic measurement
- Composite fabrication and AM

Collaboration between aerospace manufacturers, robot manufacturers, and research institutes is required to standardize testing in this way, enabling comparison between tests obtained using different accuracy improvement methods.

Calibration

Kinematic calibration is the starting point for robot accuracy improvement. Taking a first principles approach, the first axis may be considered as a transformation $\mathbf{T}_1$ from the base coordinate system to the output of this axis. The transformation has six degrees of freedom (DoF) for rigid body motion. Each DoF may be considered as a parameter with up to three components: the nominal geometry of the link, the commanded position of the axis, and an error term. In general, five of the DoF will have only nominal and error terms while a single axis will also have a commanded position. Taking such an approach for a serial robot with i axes results in $i+1$ transformations between the base coordinate system and the mechanical interface. Solving each transformation in turn gives the position and orientation of the mechanical interface with respect to the base. This is known as calculating the direct kinematics of the chain.

Controlling a robot involves solving the inverse kinematics. The required transformation from base coordinate system to end-effector coordinate system is known; this is the commanded pose of the robot. The kinematics must calculate the required joint positions. However, there are many possible solutions to the inverse kinematics, and so the controller must identify a robot pose that will allow a smooth movement from the current joint positions. These calculations are complex [13], but transformations do not need to include error terms for every DoF since the end effector is relatively insensitive to some of them. The Denavit-Hartenberg (DH) convention is a simplification that reduces the six DoF for each joint to four *DH parameters*, making inverse kinematics easier to solve. It is used by most controllers [14].

The DH convention assumes that the z-axis is positioned along the axis of each joint and that the origin of each reference frame is positioned at the intersection of the common normal between the previous axis and the current axis. For a link with revolute joints, this normally gives the length (x position), axial (z) position, twist (A rotation), and commanded positon (C rotation). The assumptions imply that there is no deformation due to in-plane bending (y position and B rotation). Where successive axes are parallel or nearly parallel, the model does not provide a stable location for the joint origin.

The actual calibration process involves moving the robot through a large number of commanded poses and measuring the position and orientation of the end effector. A laser tracker is generally the most convenient and accurate instrument to use and was originally designed for robot calibration [10]. A laser tracker, in its original form, can give the coordinates of a point on the mechanical interface of the robot. Hybrid laser tracker systems including additional angular information can also provide rotational information to give the full six DoF.

Machine tool calibration takes a different approach in which there are no parameter-reducing notations and coordinate frames are placed at intuitive locations, for example, aligned along the axis drive. This can greatly simplify the calibration process, allowing measurement systems to be aligned with an axis so that individual errors can be isolated and providing an understandable relationship between rotations and the resulting translations of objects offset from the axis of rotation. However, machine tools have much simpler kinematic arrangements with only a single solution to the inverse kinematics.

Most six-axis serial industrial robots have encoders mounted on the drive motors [15]. Gear reductions of at least 100 are typically used, and it is hysteresis, backlash, and torsional elasticity in this drivetrain that primarily limits axis accuracy [16].

Calibration has been classified into three levels [13]:

1. **Joint level** calibration of the transducers in isolation
2. **Kinematic** calibration, normally in terms of the DH parameters
3. **Nonkinematic** calibration to identify additional parameters such as joint compliance, friction, clearance, or the dynamic model

However, these classifications are not particularly helpful since Level 3 implies some unspecified parameters in addition to the DH parameters. It is more useful to specify what these are.

A full kinematic calibration combined with compliance parameters for the stiffness of joints has been demonstrated to reduce static positioning error from several millimeters down to less than 200 micrometers [17]. With direct access to the robot controller manufacturers' proprietary calibration, methods are generally able to apply these compensations at a finer level of granularity in the movement of the robot.

ElectroImpact has taken this approach to its logical conclusion, fitting high-accuracy encoders to each joint and using feedback from these to control the robot through a Siemens 840d controller with a higher-order kinematic model, achieving a global accuracy of approximately 0.25 millimeters [18], but adding considerable expense.

Recommendations

Refitting a robot with joint output encoders and a machine tool controller greatly increases the cost of a robot. Factory-fitted joint encoders that provide feedback to the robot encoder does not have to add very significant cost. However, because of the currently low demand for high-accuracy robots, high costs are passed on by the manufacturers, causing demand to remain low. Robot manufacturers should consider the potential for extensive automation of aerospace joining, AM, and composite fabrication and seek to make high-accuracy robots available at a cost that encourages this market to develop.

Datuming

ISO 9787 defines the coordinate systems for a robot [19] located within a world coordinate system $[O_0 - X_0 - Y_0 - Z_0]$

by its base coordinate system denoted $[O_1 - X_1 - Y_1 - Z_1]$ with origin O_1 as defined by the manufacturer of the robot, and the $+Z_1$ axis pointing directly upwards. The robot's mechanical interface coordinate system $[O_m - X_m - Y_m - Z_m]$ has origin O_m at the center of the mechanical interface. The $+Z_m$ axis points perpendicularly away from the mechanical interface. The accuracy of a robot can be defined as its ability to locate its mechanical interface coordinate system with respect to its base coordinate system. A kinematic model describes how the joints and links combine to achieve said accuracy [20]. There are two aspects to this, firstly calibration of a robot end effector's tool center point (TCP) and subsequently datuming the robot.

Poor datum strategy and measurement can often multiply errors a number of times. The extent of this amplification of errors is not always intuitively apparent. This has been demonstrated for industrial metrology [21, 22, 23] and is equally true for robotic positioning. Monte Carlo Simulation of datums can be used to identify the extent of this and to provide feedback for the design of improved datum strategies, as is the current state of the art for the use of coordinate measurement machines (CMMs) [24].

Recommendations

The effect of TCP calibration and datum strategy on the accuracy of robotic processes should be properly evaluated and this information used to refine datum strategies, in order to reduce this effect to acceptable levels. Monte Carlo Simulation algorithms should be included in offline programming software to enable this evaluation, as is the case for some CMM software. It is also important that consistent datum strategies are used when performing accuracy tests, so that meaningful comparisons can be made.

Adaptive Control

Adaptive control is a term used within the literature to mean the varying of control parameters, such as the kinematic parameters, during the control process [25]. It is used in this sense here. The reader should note that the term Adaptive Robotic Control has been used in a quite different sense within industry to refer to closed-loop control using an external coordinate measurement system to provide feedback on the end-effector position and orientation [26]. This form of closed-loop control is referred to in this report as *global measurement feedback* and may also be referred to as *online compensation*.

An example of the use of adaptive control for an industrial robot would involve adjusting the DH parameters to allow for thermal expansion using feedback provided by thermocouples mounted on the structure of the robot.

Work carried out at the University of Michigan 20 years ago demonstrated that there is the potential to greatly enhance the accuracy of industrial robots using more rigorous calibration, as described above, combined with adaptive control [27]. The researchers developed a control strategy in which a robot is first calibrated to identify parameters while operating at normal temperature. An enhanced set of kinematic parameters were used which include the DH parameters plus an additional kinematic parameter for rotation about the y axis, as well as the compliance of each joint, modeled as a linear spring. The calibration involved measuring the coordinates of the end effector in many poses and then solving simultaneous kinematic equations which include all of these model parameters. The researchers also discussed modeling the other DoF for the joint compliance but decided to only include the torsional stiffness about the actuated axis for each joint. Following this initial calibration, temperature sensors were positioned at multiple locations on the robot structure, and the coordinates of the end effector were measured again in multiple poses, this time while the robot was heated. These observations were used to fit an empirical model describing the temperature dependence for the DH parameters. Robots are built using different materials, with different coefficients of thermal expansion (CTE); this makes it difficult to predict thermal deformation [28]. Fitted empirical models avoid this difficulty. The thermal changes are treated as quasi-static and the control parameters updated periodically between operations, based on temperature measurements and the empirical model for temperature. This approach reduced the single-point positioning accuracy for an unspecified industrial robot from 1.2 millimeters to 0.11 millimeters.

A number of studies [15, 27] have shown that the most significant parameters can be clearly ordered in terms of their effect on robot accuracy:

1. The DH kinematic parameters
2. The additional kinematic parameter for rotation about the y-axis
3. Torsional joint compliance and backlash about the z-axis. Link and joint compliance contributes up to 10% of the position errors [29]
4. Thermal expansion and shape changes due to variation in the temperature

There are two approaches to implementing adaptive control. The first is to actually update the kinematic parameters on the robot controller. This approach depends on having access to them, which many robot manufacturers do not provide. The second approach is to use offline programming software to apply the correction to the robot path [30].

Recommendations

Significant efforts should be directed toward replicating the adaptive control work carried out by the University of Michigan in the late 1990s since they apparently achieve better accuracy than much more expensive systems developed more recently. This approach has received very little attention despite the potential to achieve high accuracy with very little hardware cost. Once the required algorithms are fully understood, only low-cost temperature sensors need to be added to the robot. Implementation must involve robot manufacturers as integration with the robot's native controller is required for a low-cost solution.

Although robot manufacturers are increasingly providing the ability to update robot positions at high frequency, this usually does not provide sufficient access to the controller to implement adaptive control. Providing more open interfaces would enable research institutes to develop more effective control strategies for high-accuracy operation.

Feedforward Control

Feedforward control involves using expected errors to apply corrections in advance of the error actually occurring. At its simplest, this can be applied to a robot that is repeating the same trajectory, with previously observed errors used to apply corrections so that successive iterations become progressively more accurate [31].

More sophisticated feedforward control algorithms may use model-based predictions to correct for machining process forces [32] and hysteresis effects such as backlash [33]. The latest work in this area shows that it is possible to implement effective backlash compensation on industrial robots [34].

Recommendations

Model-based feedforward control should continue to be developed to compensate for hysteresis effects, such as backlash, and process forces, such as gravity loads and cutting forces. This should seek to create an integrated control strategy for low-frequency and high-amplitude effects using a combination of calibration of all kinematic parameters and joint stiffness, adaptive control with thermal sensors and empirical models of thermal expansion used to update kinematic parameters, and feedforward control used to effectively compensate for joint stiffness when changing direction or loading condition.

Although robot manufacturers are increasingly providing the ability to update robot positions at high frequency, this usually does not provide sufficient access to the controller to effectively implement feedforward control. Providing more open interfaces would enable research institutes to develop increasingly effective control strategies for high-accuracy operation.

Global Measurement Feedback

The use of a measurement system to provide feedback on the position, and possibly also orientation, of the robot end effector with respect to the robot's base coordinate system is referred to here as global measurement feedback, but may also be called *online compensation* within industry.

Although calibration and adaptive control can greatly improve the static positioning accuracy of robots, in real applications it often proves to be inadequate [35]. Often parameters that are difficult to calibrate, such as backlash in axis 1, become dominant sources of error. In order to overcome the limitations of the offline calibration, it becomes necessary to directly correct the robot motion using metrology instruments in real time [36, 37]. This generally involves placing a target on the end effector which can be tracked using a metrology instrument and a computer that generates the correction commands before sending them to the robot controller. Global measurement feedback may give greater accuracy than adaptive control, but the complex systems are considerably more expensive [38].

Influences, or error sources, affecting robot accuracy may be considered as static, quasi-static, or dynamic. Static influences remain stable for long periods of time and are well suited to calibration. Quasi-static influences may change significantly over a period of hours but remain sufficiently stable that adjustment is not required during a single movement of the robot (e.g., a pass of a cutting tool or drilling a single hole). Quasi-static influences are well suited to adaptive control. Dynamic influences change rapidly. They may be corrected using adaptive control if sufficient integration with the robot controller is available. However, forms of online compensation are often required.

The method and implementation of dynamic corrections depend strongly on the manufacturer and model.

Turnkey solutions are available that provide full six DoF feedback to the robot controller from a hybrid laser-tracker system. This is mature and accurate with high-frequency feedback, the typical positioning accuracy for hole drilling applications is 0.05 millimeters [39]. It is expensive and has not until recently been applied to real-time robotic machining although turnkey solutions may be available following the Metrology-Assisted Robotic Automation project at the UK's Manufacturing Technology Center. This project demonstrated dynamic path following an accuracy of 0.2 millimeters but did not have the vibration of machining to contend with. However, it is important to consider that dynamic compensation is limited by inertial effects within the robot and not by the frequency of feedback. With a laser tracker only, a single object can be tracked and the laser must maintain a single line of sight (LOS).

Several approaches to the use of photogrammetry for six DoF tracking are possible. Fixed stereo camera systems, ad hoc networks of separate cameras, and fixed triplanar cameras are all used. For any photography-based method, there is an inherent compromise between the frequency with which images can be obtained and the resolution of those images. Any photographer will be familiar with the improved sharpness that can be achieved with a long exposure, but that for moving objects a short exposure must be used which will compromise this sharpness. This is due to signal-to-noise effects inherent in the properties of light. With digital camera systems, there may however be a restriction in capture rates imposed by data transfer before these optical limits have been reached. For these reasons, normal photogrammetry systems, which use 2D camera sensors, are not able to give the required accuracy and frequency of feedback for robotic control. Fixed triplanar cameras, in which each camera consists of a one-dimensional charge-coupled device array, are used. Triplanar cameras have some advantages in this respect as linear charge-coupled device arrays can more readily operate at simultaneously high resolution and frequency. The speed of these systems is further enhanced as they enable a direct calculation of coordinates (three angles giving three coordinates) with no bundle adjustment algorithm required. However, matching the accuracy and speed of a laser tracker is not normally possible. The main advantage of this approach is being able to track multiple targets simultaneously. This can allow the robot to be positioned relative to multiple datums on a flexible part. It can also allow tracking of the robot end effector to seamlessly switch between targets on multiples faces of the end effector as it rotates, which would cause the LOS to a laser tracker to be lost. Triplanar camera systems can achieve positioning accuracy of 200 micrometers [40]. This is also a very expensive solution.

The University of Bath has a demonstration robotic machining cell which uses only positional feedback from a standard laser tracker. It has been found that for most applications the rotational accuracy of the robot is sufficient, and significant benefits can be realized with only positional feedback. An advantage of this approach is that it can achieve an accuracy comparable to the six DoF systems but at a fraction of the cost, potentially as low as $130,000 for a robotic machining cell with integrated laser tracker.

Recommendations

Robot integrators and researchers should consider the role of global measurement feedback within an overall control strategy including calibration, adaptive control, and feedforward control. Although it may be relatively straightforward to implement a turnkey solution involving high-cost laser metrology, this does not necessarily solve issues of dynamic control, and it may cause lock-in to high-cost hardware. More sophisticated control solutions that compensate for errors at their source are likely to result in more appropriate solutions for many applications, albeit with considerably more complex development required. The possibility of using only positional feedback from lower-cost laser tracker systems should also be considered within this overall strategy.

Local Measurement Feedback

Where a robot is required to position or align with respect to a local feature, it often makes little sense to rely on the global accuracy of the robot. Instead a local measurement or vision system can be used to measure the position of the end effector with respect to the relevant feature, and provide feedback to the robot controller. Such local sensors are often low cost and can provide highly accurate feedback. Force transducers may be included within this area, effectively turning the end effector or component being assembled into a measurement probe that feeds back to align the robot.

An example of where local measurement feedback can provide an ideal solution is drilling and fastener insertion. Consider a case where two robots are working collaboratively to perform this process. The first robot drills holes while the second robot inserts fasteners into these holes. Without any local measurement feedback, these robots would need to be located to a common datum structure and maintain positional accuracy of perhaps 50 micrometers. This would be extremely challenging and essentially not possible with current technology. However, if the second robot is fitted with a relatively low-cost laser line measurement system, it could easily locate the position of the hole to the required accuracy for fastener insertion. In order to maintain alignment of the fastener during insertion, two axes force transducers in the end effector could be used to provide further feedback. Such an approach has been demonstrated to enable solid rivets to be installed using robots approaching the hole from each side of the panel [41]. There are many different applications for local measurements and vision-based systems [42]. A monocular has been used to align the end effector to within 0.2 degrees [43].

Recommendations

Relatively simple laser line, contact probe, or force feedback systems should be considered before implementing highly complex control to improve robot accuracy. Although these technologies are relatively mature, industry still requires a clearly defined "tool-box" from which the most appropriate methods that can be selected for any given process.

Locally Clamped Machine Tools

An alternative approach to achieving high-accuracy drilling and machining using robots is to use a localized drilling head that can attach itself to multiple zones on the product using some form of local clamping such as vacuum clamping. Datuming may be carried out using a laser tracker or photogrammetry. This can give stiffness and accuracy comparable with a machine tool but operating over a small volume. The robot then effectively becomes just a positioning device to move a small machine tool around on the part.

This approach is suitable for use with both articulated robots employed to move a local machining head between

FIGURE 4. **Flextrack robot joining fuselage panels longitudinally for Airbus A320 assembly; courtesy of Airbus.**

zones, or a crawling robot that moves continuously over a panel [44]. An example is the *Flextrack* robots used on the recently opened Airbus A320 fuselage assembly plant in Hamburg. In both cases, a single 3D laser tracker may be used to datum several robots, since they only need to be datumed after clamping onto a machining zone, and do not require feedback from the laser tracker while operating within this zone. Once clamped, the machining head could be used to perform orbital drilling as well as fettling operations (Figure 4).

Recommendations

The potential for locally clamped machine tools to provide low-cost automation should be explored further through industrial trials.

Reducing Cutting Disturbances

The determination of cutting parameters, such as spindle speed, cutter type, feed rate, number of passes (one shot or two shots), and coolant, always involves a trade-off between considerations such as surface finish and material removal rate. Machine tools have high stiffness as a primary design consideration and are many times stiffer than industrial robots. The resonance of a robot is also highly dependent on its pose, making the tuning of cutting speed to avoid natural resonances more difficult. The trade-offs when cutting metal, whether drilling or surface milling, using a robot will, therefore, involve considerably lower material removal rates.

Conventional drilling has issues with high on-axis reaction forces and levels of vibration. This makes it difficult for the relatively flexible structures of industrial robots to produce good-quality, accurate holes. Holes can also be produced by interpolating a circular path using a milling machine with a smaller diameter cutting tool, known as orbital drilling [45]. This results in lower reaction forces and lower vibration. However, heavy machine tools able to drill holes in large airframes are not flexible or reconfigurable enough for lean production systems. Conversely, industrial robots can't interpolate circles accurately enough due to the combination of stiffness and inertial effects, as well as backlash and servo mismatch. Orbital drilling provides a method of interpolating machined holes using flexible automation [46]. It effectively uses a very small machine tool that has just enough movement along each axis to interpolate a hole. This orbital drilling machine is then positioned where the whole is required in the same way that a conventional drilling machine would be positioned. The machine is light enough for a robot to handle it, enabling flexible and reconfigurable drilling within large assemblies.

There is a parallel between using orbital drilling to reduce drilling forces and making smaller passes when machining. In both cases, the disturbing reaction forces caused by the cutting tool are reduced, enabling a more flexible automation system to produce high-accuracy features, albeit at lower speed.

Recommendations

Work to optimize cutting parameters should be carried out in conjunction with the development of highly dynamic end effectors, to reduce the effects of vibration and chatter on machined surfaces.

Highly Dynamic End Effectors

The limiting factor for the frequency of dynamic control is currently the inherent inertial effects of the robot [47]. This means that any improvement in measurement data or communication rate between the metrology system and the robot will not improve accuracy, since the robot cannot physically accelerate quickly enough to respond to this feedback. To achieve even higher dynamics than the robot/controller can provide, highly dynamic end effectors have been developed. These are relatively small motion stages that have faster response times and can work independently from the robot controller [48, 49, 50]. The University of Bath is developing a hybrid approach in which highly dynamic end effectors are used to compensate high-frequency low-amplitude errors (vibration) while a laser tracker providing feedback to the robot controller is used to compensate low-frequency high-amplitude (path following) errors. This is envisaged to be a system which is considerably lower in cost than current six DoF feedback systems while offering significantly improved accuracy for highly dynamic applications such as machining.

For very high-speed and low-amplitude disturbances, such as vibration, the mechanical dynamics of the robot structure and drives limits the ability to apply corrections. Localized active damping systems may therefore be required. Current methods differentiate between low-frequency high-amplitude disturbances, which are corrected by providing feedback to the robot controller, and high-frequency low-amplitude disturbances. Two approaches have been taken to apply the high-frequency corrections.

The first approach is to directly control the spindle positon on the end of the robot structure using small-motion stages driven by piezo actuators [51]. The second approach is using actuators attached to a reaction mass to correct the position of the robot end effector.

Recommendations

High dynamic compensation, or active vibration control, is a requirement for high-accuracy machining using a robot. Further efforts are required to perfect the control strategy. These methods must be combined with low-frequency and high-amplitude compensation systems to achieve global accuracy.

Hybrid High-Accuracy Control

Typical industrial robot path-following accuracy is a few millimeters, although some off-the-shelf high accuracy models are able to achieve approximately 0.5 millimeters. This is achieved through calibration of kinematic parameters and joint stiffness as well as potentially placing encoders closer to joints to lesson backlash and joint flexibility issues. Studies on the significance of the various influences on robot accuracy indicate that the most significant influences are, starting with the most significant, 1) the DH kinematic parameters, 2) the kinematic parameter missing from the DH convention for rotation about the y axis, 3) torsional joint compliance and backlash about the z axis, and 4) thermal expansion and shape changes due to variation in the temperature. The first two influences are fully corrected in such a case. The third influence may also be corrected. The fourth influence, thermal effects, is not thought to be corrected in any commercially available system. Research-based adaptive control in which kinematic parameters are adjusted for thermal expansion shows potential for accuracy as low as 0.1 millimeters from a calibrated robot with no external positional feedback. State of the art for commercially available robot accuracy can typically be able to achieve 0.05 millimeters accuracy for drilling, although for dynamic applications this is reduced to 0.2 millimeters.

Requirements for high accuracy and human-robot collaboration lead to integrated robotic systems that are many times more expensive than the robots themselves. This is largely due to the fact that the robots were not designed for such applications, and expensive metrology instruments are integrated to improve their performance. The number of robots currently required for aerospace drilling and fettling operations makes the developments of specialized robots uneconomic. However, there are many other potential applications in aerospace such as AM, composite fabrication, and inspection. Furthermore, there are many applications within automotive manufacturing that have not yet been automated due to insufficient accuracy in current robotic systems, and future automotive production is expected to require even higher accuracy and more rigid robots [52].

Recommendations

Development of improved accuracy robots must consider several possible strategies. For each strategy, the achievable static positioning accuracy, dynamic path following accuracy, and dynamic response to disturbances must be considered, along with the cost. It will then be possible to identify which accuracy improvement strategy is most appropriate.

For low-frequency, high-amplitude disturbances there are essentially three strategies that could be employed:

1. Some combination of improved robot calibration, datuming, adaptive control, and feedforward control
2. Global metrology feedback
3. Local metrology feedback

Where local metrology feedback can be applied, this will probably be the preferred approach. Global metrology feedback currently offers the highest accuracy from an off-the-self solution, and further work is required to understand the scope of accuracy improvement achievable through improved control of the robot itself.

For high-frequency, low-amplitude disturbances, as encountered when machining or drilling, active vibration control using accelerometers and reaction masses provides a promising technology but requires further development to provide an industrial solution.

Improved access to robot controllers, particularly the ability to update kinematic parameters in real time would greatly enhance the development of improved adaptive control schemes. It appears that there may be a considerable scope for improvements through adaptive control that uses dynamic models for joint compliance, together with temperature sensors to correct for thermal expansion using empirical models.

Greater emphasis needs to be given to developing more cost-effective solutions. The cost of integrated high-accuracy robot cells is currently over $1 million.

Human-Robot Collaboration

A final generic unresolved area, not related to improving robot accuracy, is improved collaboration with humans. This can be seen as happening in two phases. The first is the collaborative workspace in which robots and humans perform separate and discrete tasks, but can work safely within the same space. This requires increased redundancy within robot-control hardware and software, for example, using two-channel measurement systems with fail-safe bus and input/output systems. It may also be possible to use robot models already running in real time within robot controllers to detect faults. There are two approaches currently taken to enable safe collaborative workspaces. The first is to use standard robots with interlocked guarding replaced by light guards or camera systems [53]. These do not necessarily stop the robots completely but may be set to ensure that, as people get closer to the robots, they slow down to a safe working speed. The other approach in this phase is to use a purpose-built collaborative robot, or "cobot," that incorporates torque sensors within its joints to prevent it from crushing a person.

The second phase in human-robot collaboration is collaborative robotics [54] in which humans and robots work together on the same tasks. This might involve a robot passing tools to a human worker, a bit like a robotic dental nurse passing tools to a human dentist, or it might involve a robot performing laborious or dangerous tasks that require human guidance, such as applying sealant over complex and variable geometries. Collaborative robotics may require new ways of controlling robots, such as gesture control [55] or voice command.

Recommendations

Human-robot collaboration is an emerging technology, and it is difficult to predict how it will develop. A key area for aerospace manufacturing is identifying which processes to automate and which are best carried out by human workers. This does not necessarily mean that robots should be performing the hard operations such as cutting and drilling, they may prove of more use in supporting roles such as logistics and tool handling.

Conclusions

Improving accuracy is a generic requirement that will enable operations such as machining, AM, composite fabrication, drilling, automated program development, and inspection. It is perhaps the single biggest challenge preventing widespread adoption of robots in aerospace manufacturing. There is a need to define standards for robot accuracy that are relevant to process requirements and enable comparison between both different commercially available robots and experimental results.

Most robot accuracy improvement should be addressed by directly compensating for the errors present during operation. The optimum approach should be determined considering the calibration of all kinematic and joint stiffness parameters, the use of joint output encoders, adaptive control including empirical models for thermal expansion, and feedforward control that compensates for hysteresis and external loads. There is scope to improve calibration-based robotic control by including additional kinematic parameters and joint stiffness parameters, and by using joint output encoders. It is not currently clear how best to combine these methods with others, such as adaptive control and feedforward control. Adaptive control has been shown to dramatically increase robot accuracy to levels better than achieved using most forms of external metrology feedback. This is done by updating kinematic parameters using temperature sensors and empirical models for thermal expansion. Feedforward control provides the best means of compensating for hysteresis effects such as backlash and also for low-frequency force disturbances such as gravity loads and some process forces. The impact of datuming on robot accuracy is currently not properly considered in process planning or most research and development. Monte Carlo Simulation could be used to model and optimize these effects.

Global measurement feedback is currently the highest-accuracy turnkey solution, but it is costly and has limited capability to compensate dynamic errors. There is potential for more cost-effective systems using 3D laser trackers combined with other methods. Local measurement feedback is a mature, affordable, and highly accurate technology where the robot is required to position or align relative to some local feature, for example, inserting a fastener into a hole or applying sealant along an edge. Locally clamped machine tools are an alternative approach that can utilize the flexibility of industrial robots while also enabling high-quality machined surfaces.

Highly dynamic end effectors measure high-frequency disturbances using inertial sensors and provide feedback directly to the end effector using a reaction mass. This is the only technology able to compensate for high-frequency disturbances and must be combined with other accuracy enhancements for high-quality robotic machining. Reducing cutting disturbances through a better understanding of how to optimize cutting parameters for the specific process will be an important part of realizing robotic machining.

Hybrid high-accuracy control strategies will be required for many processes. For example, the highest accuracy dynamic control is likely to involve a combination of enhanced calibration, adaptive control, feedforward control, and highly dynamic end effector.

Human-robot collaboration is an emerging area that is hard to predict, but it is clear that aerospace manufacturing needs to understand which processes to automate and which are best carried out by human workers.

SAE EDGE™ Research Reports

SAE EDGE™ Research Reports, like the present report on "Unsettled Technology Domains in Robotics for Automation in Aerospace Manufacturing," are intended to push further out into still unsettled areas of technology of interest to the mobility industry. SAE launches these reports before attempting to form a joint working group, let alone a CRP or a Standards Committee.

SAE EDGE™ reports are intended to be quick, concise overviews of major unsettled areas where vital new technologies are emerging. By definition, an unsettled area is characterized more by confusion and controversy than established order. Early practitioners must confront an absence of agreement. Their challenge is often not to seize the high ground but to find common ground. These scouting reports from the frontiers of investigation are intended merely to begin the process of sorting through critical issues, contributing to a better understanding of key problems, and providing helpful suggestions about possible next steps and avenues of investigation.

SAE EDGE™ Research Reports, therefore, are fundamentally distinct from the more formal working groups approach and far removed from the more mature CRP and standards development process.

Next Steps

This publication should be considered only as a first step toward clarifying the issues around robotics and automation in aerospace manufacturing. The intention behind this and other SAE EDGE™ Research Reports is to start a dialogue among interested parties on important industry-wide topics that require further attention. The expectation is that these explorations of unsettled areas of technology will lead to the formation of working groups and, ultimately, committees that can address and resolve the issues they raise, producing a framework for developing a common vocabulary of definitions, best practices, protocols, and standards needed to support continued progress toward safer and more innovative automotive products.

The experts' meeting that gave rise to this publication demonstrated a great willingness on the part of the industry to define the terminology, procedures, and eventually the standards needed to enable robotic aerospace manufacturing to move ahead as quickly and efficiently as possible.

This SAE EDGE™ Research Report on robotics and automation in aerospace manufacturing identifies the following key topics for further pursuit, both through continued informal discussions among industry practitioners and through more formal working groups:

- Define standards for robot accuracy to improve calibration-based robotic control
- Understand the impact of **datuming** on robot accuracy
- Adaptive control
- Feedforward control
- Global measurement feedback
- Local measurement feedback
- Locally clamped machine tools
- Reducing cutting disturbances
- Highly dynamic end effectors
- Hybrid high-accuracy control
- Human-robot collaboration

Recommendations

- There is a need to **define standards for robot accuracy** that are relevant to process requirements and enable comparison between both different commercially available robots and experimental results.
- There is scope to improve **calibration**-based robotic control by including additional kinematic parameters and joint stiffness parameters, and by using joint output encoders. It is not currently clear how best to combine these methods with other methods, such as adaptive control and feedforward control.

- The impact of **datuming** on robot accuracy is currently not properly considered in process planning or most research and development. Monte Carlo Simulation could be used to model and optimize these effects.
- **Adaptive control** has been shown to dramatically increase robot accuracy to levels better than achieved using most forms of external metrology feedback, by updating kinematic parameters using temperature sensors and empirical models for thermal expansion.
- **Feedforward control** provides the best means of compensating for hysteresis effects such as backlash and also for low-frequency force disturbances such as gravity loads and some process forces.
- **Global measurement feedback** is currently the highest-accuracy turnkey solution, but it is costly and has limited capability to compensate dynamic errors. There is potential for more cost-effective systems using 3D laser trackers combined with other methods.
- **Local measurement feedback** is a mature, affordable, and highly accurate technology where the robot is required to position or align relative to some local feature.
- **Locally clamped machine tools** are an alternative approach that can utilize the flexibility of industrial robots while also enabling high-quality machined surfaces.
- **Reducing cutting disturbances** through a better understanding of how to optimize cutting parameters for the specific process will be an important part of realizing robotic machining.
- **Highly dynamic end effectors** measure high-frequency disturbances using inertial sensors and provide feedback directly to the end effector using a reaction mass. This is the only technology able to compensate for high-frequency disturbances and must be combined with other accuracy enhancements for high-quality robotic machining.
- **Hybrid high-accuracy control** strategies will be required for many processes. For example, the highest accuracy dynamic control is likely to involve a combination of enhanced calibration, adaptive control, feedforward control, and highly dynamic end effector.
- **Human-robot collaboration** is an emerging area that is hard to predict, but it is clear that aerospace manufacturing needs to understand which processes to automate and which are best carried out by human workers

Highly dynamic end effectors should be used where the frequency of disturbances exceeds the dynamic control capability of a robot.

Abbreviations/Definitions

2-D - Two-dimensional

3-D - Three-dimensional

AGV - Automated guided vehicle

AM - Additive manufacturing

CMM - Coordinate measurement machine

CNC - Computer Numerical Control

CTE - Coefficient of thermal expansion

DED - Directed Energy Deposition

DfM - Design for manufacture

DH - Denavit-Hartenberg

DoF - Degrees of Freedom

iGPS - Indoor Global positioning System

LOS - Line of sight

PKM - Parallel kinematic machine

TCP - Tool center point

Acknowledgments

Recognition should go first to all the participants, many of who also provided feedback on the draft version of this publication. Without their input and initiative, this SAE EDGE™ Research Report would not have been possible.

Dave Burns, Kuka Systems
Dr. Andrew Francis, GKN Aerospace
Russell DeVlieg, ElectroImpact
Roger Holden, True Position Robotics
Dr. Jim Heley, Advanced Aerospace Assembly Ltd
Andy Silcox, AMRC, Broughton
Dr. Zheng Wang, University of Bath

The authors of this document together with the SAE Team responsible for its creation join in expressing our deepest appreciation to all the individuals mentioned above.

Dr. Jody E. Muelaner
Muelaner Engineering Ltd.

Bibliographic References

1. Airbus Press Office, "A320 Family: The Most Successful Aircraft Family Ever," in *Facts & Figures,* 2019, Airbus, Doc.

2. "Boeing to cut 737 Max aircraft production by nearly 20%," in *The Guardian,* 2019.
3. Airbus, "Global Market Forecast 2007-2026," 2007, Doc
4. Muelaner, J.E., Wang, Z., Martin, O., Jamshidi, J. et al., "Estimation of Uncertainty in Three Dimensional Coordinate Measurement by Comparison with Calibrated Points," *Measurement Science and Technology* 21(2):9, 2010, http://dx.doi.org/10.1088/0957-0233/21/2/025106.
5. Barnes, J., Rizos, C., and Wang, J., "Locata: The Positioning Technology of the Future?," in *6th International Symposium on Satellite Navigation Technology Including Mobile Positioning & Location Services*, Melbourne, Australia, 2003.
6. Priyantha, N.B., *The Cricket Indoor Location System* (Massachusetts Institute of Technology, 2005).
7. Mautz, R., *The Challenges of Indoor Environments and Specification on Some Alternative Positioning Systems* (Hannover, Germany: IEEE Computer Society, 2009).
8. ISO 9283, "Manipulating Industrial Robots," Performance criteria and related test methods, 1998.
9. Nubiola, A., Slamani, M., Joubair, A., and Bonev, I.A., "Comparison of Two Calibration Methods for a Small Industrial Robot Based on an Optical CMM and a Laser Tracker," *Robotica* 32(3):447-466, 2014, http://dx.doi.org/10.1017/s0263574713000714.
10. Lau, K., Hocken, R., and Haight, W., *Automatic Laser Tracking Interferometer System for Robot Metrology* (Switz: Interlaken, 1985).
11. Jozwik, J., Kuric, I., Ostrowski, D., and Dziedzic, K., "Industrial Robot Accuracy Testing with QC20-W Ballbar Diagnostic System," *Manufacturing Technology* 16(3):519-524, 2016.
12. Nubiola, A. and Bonev, I.A., "Absolute Robot Calibration with a Single Telescoping Ballbar," *Precision Engineering* 38(3):472-480, 2014, http://dx.doi.org/10.1016/j.precisioneng.2014.01.001.
13. Roth, Z.S., Mooring, B.W., and Ravani, B., "Overview of Robot Calibration," *IEEE Journal of Robotics and Automation* RA-3(5):377-385, 1987.
14. Sciavicco, L. and Siciliano, B., *Modelling and Control of Robot Manipulators* 2nd Edition (Springer, 2000).
15. Slamani, M., Nubiola, A., and Bonev, I., "Assessment of the Positioning Performance of an Industrial Robot," *Industrial Robot* 39(1):57-68, 2012, http://dx.doi.org/10.1108/01439911211192501.
16. Schneider, U., Ansaloni, M., Drust, M., Leali, F. et al., *Experimental Investigation of Sources of Error in Robot Machining* (Porto, Portugal: Springer Verlag, 2013).
17. Nubiola, A., Slamani, M., Joubair, A., and Bonev, I.A., "Comparison of Two Calibration Methods for a Small Industrial Robot Based on an Optical CMM and a Laser Tracker," *Robotica* 32(3):447-466, 2014.
18. Devlieg, R., "Expanding the Use of Robotics in Airframe Assembly via Accurate Robot Technology," *SAE Int. J. Aerosp.* 3(1):198-203, 2010. https://doi.org/10.4271/2010-01-1846.
19. ISO 9787, "Robots and Robotic Devices-Coordinate Systems and Motion Nomenclatures," 2013.
20. Everett, L., Driels, M., and Mooring, B., "Kinematic Modelling for Robot Calibration," in *1987 IEEE Int. Conf. Robot. Autom Proceedings*, Raleigh, NC, 1987, 183-189.
21. Forbes, A.B. and Harris, P.M., "Simulated Instruments and Uncertainty Estimation," *From the Software Support for Metrology Programme* (London: NPL, 2000), Doc Ref CMSC 01/00.
22. Djurdjanovic, D. and Ni, J., "Stream-of-Variation (SoV)-Based Measurement Scheme Analysis in Multistation Machining Systems," *IEEE Transactions on Automation Science and Engineering* 3(4):407-422, 2006.
23. Muelaner, J.E., Wang, Z., Keogh, P.S., Brownell, J. et al., "Uncertainty of Measurement for Large Product Verification: Evaluation of a Large Aero Gas Turbine Engine," *Meas. Sci. Technol* 27(11):12, 2016.
24. Waldele, F. and Schwenke, H., "Automated Calculation of Measurement Uncertainties on CMMs - Towards Industrial Application Automatische Bestimmung der Messunsicherheiten auf KMGs auf dem Weg in die industrielle Praxis," *Technisches Messen* 69(12):550-557, 2002.
25. Lewis, R.A., "Adaptive Control of a Robotic Manipulator," in *Proceedings of the IEEE Conference on Decision and Control*, New Orleans, LA, 1977, 743-748.
26. Summers, M.D., Green, B.J., and Holden, R., "Program-Controlled Process," patent number: EP1899774 (A1), 2008.
27. Gong, C., Yuan, J., and Ni, J., "Nongeometric Error Identification and Compensation for Robotic System by Inverse Calibration," *International Journal of Machine Tools and Manufacture* 40(14):2119-2137, 2000, http://dx.doi.org/10.1016/s0890-6955(00)00023-7.
28. Heisel, U., Richter, F., and Wurst, K.H., "Thermal Behavior of Industrial Robots and Possibilities for Errors Compensation," *CIRP Annals - Manufacturing Technology* 46:283-286, 1997.
29. Mustafa, S.K., Tao, P.Y., Yang, G., and Chen, I.M., *A Geometrical Approach for Online Error Compensation of Industrial Manipulators* (Institute of Electrical and Electronics Engineers Inc, 2010).
30. RoboDK, "Simulate Robot Applications," 2019. www.robodk.com [cited.]
31. Atkeson, C.G. and McIntyre, J., "An Application of Adaptive Feedforward Control to Robotics," . In: *Adaptive Systems in Control and Signal Processing.* (Sweden, Lund, 1986), 283-288.

32. Grundel, L., Lienenluke, L., Storms, S., and Brecher, C., *Compensation of Process Forces with a Model-Based Feed-Forward Control for Robot Machining* (Nanjing, China: Institute of Electrical and Electronics Engineers Inc, 2019).

33. Cordes, M. and Hintze, W., "Offline Simulation of Path Deviation due to Joint Compliance and Hysteresis for Robot Machining," *International Journal of Advanced Manufacturing Technology* 90(1-4):1075-1083, 2017, http://dx.doi.org/10.1007/s00170-016-9461-z.

34. Yamada, S. and Fujimoto, H., "Precise Joint Torque Control Method for Two-Inertia System with Backlash Using Load-Side Encoder," *IEEJ Journal of Industry Applications* 8(1):75-83, 2019, http://dx.doi.org/10.1541/ieejjia.8.75.

35. Summers, M., "Robot Capability Test and Development of Industrial Robot Positioning System for the Aerospace Industry," . In: *AeroTech Congress & Exhibition, Grapevine.* (Texas, SAE International, 2005), 1-13.

36. Kihlman, H., Loser, R., Cooke, A., Sunnanbo, A. et al., "Metrology-integrated Industrial Robots - Calibration, Implementation and Testing," in *Proceedings of the 35th International Symposium on Robotics*, Paris, France, 2004.

37. Holden, R., Lightowler, P., and Andreou, S., "Robot Accuracy: Online Compensation (EU COMET Project)," *SAE Int. J. Aerosp* 7(2):274-279, 2014, https://doi.org/10.4271/2014-01-2257.

38. Droll, S., "Real Time Path Correction of Industrial Robots with Direct End-Effector Feedback from a Laser Tracker," *SAE Int. J. Aerosp.* 7(2):222-228, 2014, https://doi.org/10.4271/2014-01-2248.

39. Leica, "PCMM System Specifications: Leica Absolute Tracker and Leica T-Products," 2008.

40. Nikon Metrology, ARC Specs, 2016 [cited 2016]; 26th Feb]. Available from: http://www.nikonmetrology.com/en_EU/Products/Large-Volume-Applications/Adaptive-Robot-Control/Adaptive-Robot-Control/(specifications).

41. Webb, P., Eastwood, S., Jayaweera, N., and Chen, Y., "Automated Aerostructure Assembly," *Industrial Robot: An International Journal* 32(5):383-387, 2005.

42. Iureva, R.A., Raskin, E.O., Komarov, I.I., Maltseva, N.K. et al., *Industrial Robot's Vision Systems* (San Francisco, CA, United States: SPIE, 2016).

43. Lei, J.-Z., Zeng, L.-B., and Ye, N., "Research on Industrial Robot Alignment Technique with Monocular Vision," *Guangxue Jingmi Gongcheng/Optics and Precision Engineering* 26(3):733-741, 2018, http://dx.doi.org/10.3788/ope.20182603.0733.

44. Collado, V., Arana, J., and Saenz, A., "A Crawling Portable Robot for Drilling Operations in Large Air Frame Components," in *AeroTech Congress & Exhibition*, Grapevine, Texas, 2005.

45. Lindqvist, R., Eriksson, I., and Wolf, M., *Orbital Drilling of Sandwich Constructions for Space Applications* (Seattle, WA, United States: SAE International, 2001).

46. Kihlman, H., Eriksson, I., and Ennis, M., *Robotic Orbital Drilling of Structures for Aerospace Applications* (Chester, United Kingdom: SAE International, 2002).

47. Abele, E., Weigold, M., and Rothenbucher, S., "Modeling and Identification of an Industrial Robot for Machining Applications," *CIRP Annals - Manufacturing Technology* 56(1):387-390, 2007, http://dx.doi.org/10.1016/j.cirp.2007.05.090.

48. Sörnmo, O., Olofsson B., Schneider U., Robertsson A., and Johansson R.. "Increasing the Milling Accuracy for Industrial Robots Using a Piezo-Actuated High-Dynamic Micro Manipulator". in *IEEE/ASME International Conference on Advanced Intelligent Mechatronics (AIM)*, Kachsiung, 2012, 104-110

49. Loser, R. and Kleinkes, M., "Real-Time Robot Positioning based on Measurement Feedback Control," *SAE Int. J. Mater. Manf.* 9(1):106-111, 2016, https://doi.org/10.4271/2015-01-2617.

50. Lehmann, C., Pellicciari, M., Drust, M., and Gunnink, J.W., *Machining with Industrial Robots: The COMET Project Approach* (Porto, Portugal: Springer Verlag, 2013).

51. Schneider, U., Drust, M., Puzik, A., and Verl, A., *Compensation of Errors in Robot Machining with a Parallel 3D-Piezo Compensation Mechanism* (Setubal, Portugal: Elsevier B.V, 2013).

52. Brogardh, T., "Present and Future Robot Control Development: An Industrial Perspective," *Annual Reviews in Control* 31(1):69-79, 2007, http://dx.doi.org/10.1016/j.arcontrol.2007.01.002.

53. Woods, T., "Eye Opener," *Engineering* 248(5):55-56, 2007.

54. Schraft, R.D., Meyer, C., Parlitz, C., and Helms, E., *PowerMate - A Safe and Intuitive Robot Assistant for handling and Assembly Tasks* (Barcelona, Spain: Institute of Electrical and Electronics Engineers Inc, 2005).

55. Asif, S. and Webb, P., "Contactless Medium Scale Industrial Robot Collaboration," *Advances in Robotics & Automation* 5(1), 2016, doi:10.4172/2168-9695.1000145.

Contact Information

EDGEresearch@sae.org.

The authors of this document together with the SAE Team responsible for its creation join in expressing our deepest appreciation to all of the individuals who contributed.

www.ingramcontent.com/pod-product-compliance
Lightning Source LLC
LaVergne TN
LVHW070941210926
848482LV00003B/278
* 9 7 8 1 4 6 8 6 0 1 2 3 7 *